Changing Water

Katie Peters

GRL Consultants,
Diane Craig and Monica Marx,
Certified Literacy Specialists

Lerner Publications ◆ Minneapolis

Note from a GRL Consultant
This Pull Ahead leveled book has been carefully designed for beginning readers. A team of guided reading literacy experts has reviewed and leveled the book to ensure readers pull ahead and experience success.

Lerner Publications Company
A division of Lerner Publishing Group, Inc.
241 First Avenue North
Minneapolis, MN 55401 USA

For reading levels and more information, look up this title at www.lernerbooks.com.

Main body text set in Memphis Pro 24/39
Typeface provided by Linotype.

Photo Acknowledgments
The images in this book are used with the permission of: © Shutterstock, pp. 3, 4–5, 12–13, 14–15, 16 (right); © iStockphoto, pp. 6–7, 8–9, 10–11, 16 (left), 16 (center)

Front cover: © Shutterstock

Library of Congress Cataloging-in-Publication Data

Names: Peters, Katie, author.
Title: Changing water / Katie Peters.
Description: Minneapolis : Lerner Publications, [2020] | Series: Science all around me (Pull ahead readers - Nonfiction) | Includes index. | Audience: Age 4–7. | Audience: K to Grade 3.
Identifiers: LCCN 2018058167 (print) | LCCN 2019000452 (ebook) | ISBN 9781541562271 (eb pdf) | ISBN 9781541558472 (lb : alk. paper) | ISBN 9781541573321 (pb : alk. paper)
Subjects: LCSH: Matter—Properties—Juvenile literature. | Change of state (Physics)— Juvenile literature.
Classification: LCC QC173.36 (ebook) | LCC QC173.36 .P46 2020 (print) | DDC 530.4—dc23

LC record available at https://lccn.loc.gov/2018058167

Manufactured in the United States of America
1 – CG – 7/15/19

Contents

Changing Water

Here is a glass of water.
Water can change.

Water can freeze.
It changes into ice.

Ice can melt.

It changes into water.

Water can boil.

It changes into steam.

Steam can cool.
It changes into water.

Water can change into
ice and steam.

Did You See It?

ice

steam

water

Index